Mammals

D0307146

LONDON, NEW YORK,
MELBOURNE, MUNICH, and DELHI

Written and edited by Sarah Walker
and Anna Lofthouse
Designed by Jacqueline Gooden
Managing editor Susan Leonard
Managing art director Cathy Chesson
Senior editor Caroline Bingham
Jacket design Chris Drew
Picture researcher Brenda Clynch
Production Shivani Pandey
DTP designer Almudena Díaz
Consultant Nick Lindsay

REVISED EDITION
DK UK
Senior editor Caroline Stamps
Senior art editor Rachael Grady
Jacket editor Manisha Majithia
Jacket design Natasha Rees
Jacket design development manager
Sophia M Tampakopoulos Turner
Producer (print production) Mary Slater
Producer (pre-production) Rachel Ng
Publisher Andrew Macintyre

DK INDIA
Senior editor Shatarupa Chaudhuri
Senior art editor Rajnish Kashyap
Editor Surbhi Nayyar Kapoor
Art editor Shipra Jain
Managing editor Alka Thakur Hazarika
Managing art editor Romi Chakraborty
DTP designer Dheeraj Singh
Picture researcher Sumedha Chopra

First published in Great Britain in 2002
This edition published in Great Britain in 2015
by Dorling Kindersley Limited
80 Strand, London WC2R 0RL

Copyright © 2002, © 2015 Dorling Kindersley Limited
A Penguin Random House Company

13 14 15 16 17 10 9 8 7 6 5 4 3 2 1
001 – 196635 – 02/15

All rights reserved. No part of this publication may be reproduced, stored in a retrieval system, or transmitted in any form or by any means, electronic, mechanical, photocopying, recording, or otherwise, without the prior written permission of the copyright owner.

A CIP catalogue record for this book
is available from the British Library.

ISBN 978-1-4093-4428-5

Colour reproduction by Scanhouse, Malaysia
Printed and bound in China by Hung Hing

Discover more at
www.dk.com

Contents

Mammal world

You might wonder if tiny mice, huge whales, and humans have anything in common. They do – they are mammals. All mammals have hair or fur; are warm-blooded as they have a constant body temperature; and feed their young on milk.

Odd eggs out
Most mammals are born, but the hedgehog-like echidnas (right) and the duck-billed platypus hatch out from eggs.

Instant food
Almost all female mammals suckle their young on milk. The milk provides the best balance of fat and proteins so that young mammals can grow quickly.

There are more than 5,000 mammal species

Hair or fur helps to keep heat in.

LONG HAUL

Elephant mums are mammal record breakers for having the longest pregnancies – nearly two years. Young elephant calves grow inside their mothers like most other mammals (for example, human babies or kittens). When born, a calf can weigh 90 kg (200 lb). So at the end of the pregnancy, that's like a mum carrying 90 bags of sugar!

Extended family

Humans belong to a group of mammals called primates. Other primates include monkeys and apes, so they are our closest mammal relatives.

Open wide

A hippopotamus would be perfect at the dentist's with its large mouth and wide jaw stretch. All mammals have distinct jaws, where the lower jaw is hinged directly to the skull.

on Earth and the human species is just one of them.

Mammals are the only animals to have ear flaps.

Amazing mammals

Mammals come in all shapes and sizes, but those that look alike belong to the same groups, or orders. Each order contains different types, or species, of mammals.

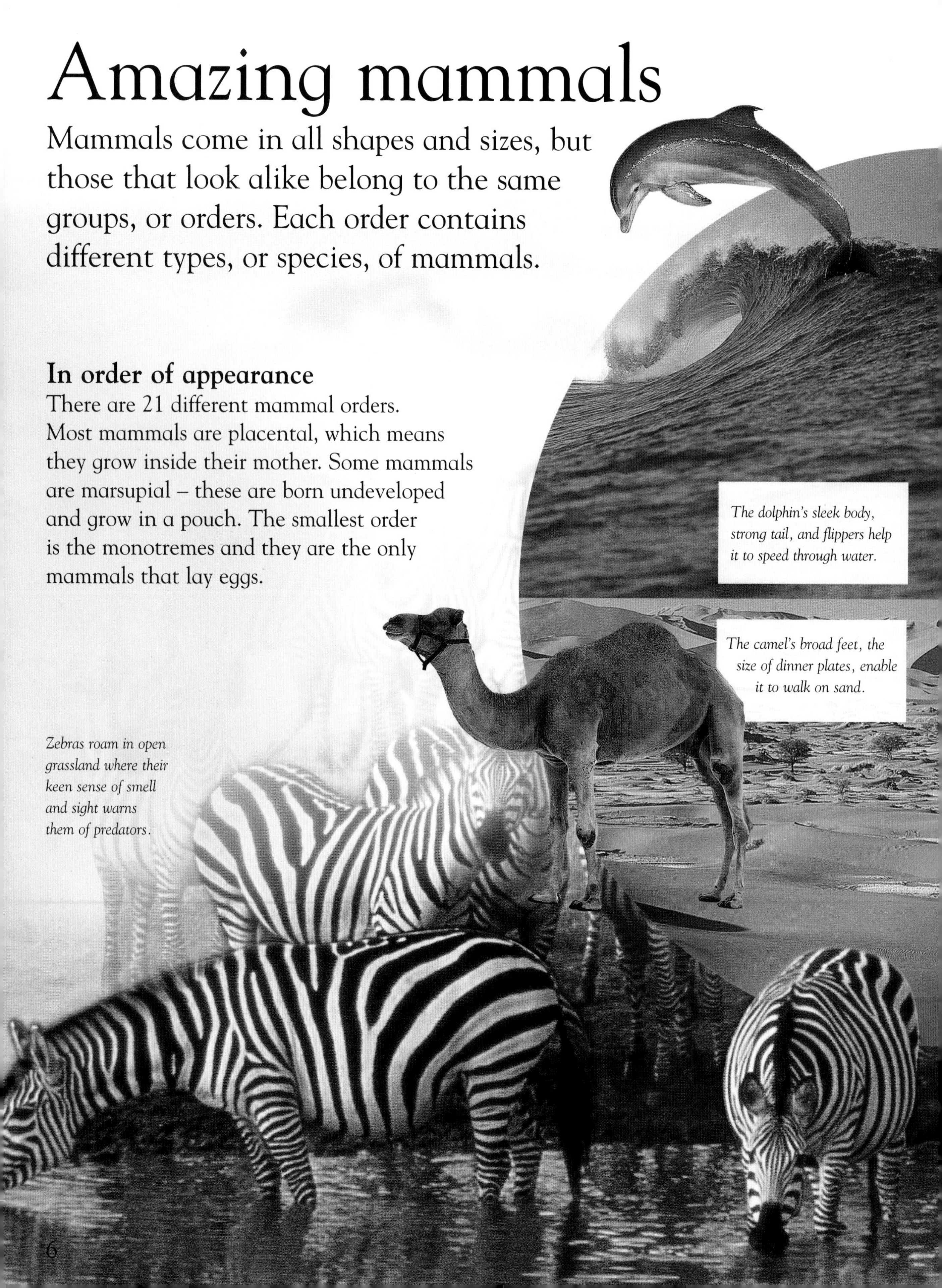

In order of appearance

There are 21 different mammal orders. Most mammals are placental, which means they grow inside their mother. Some mammals are marsupial – these are born undeveloped and grow in a pouch. The smallest order is the monotremes and they are the only mammals that lay eggs.

The dolphin's sleek body, strong tail, and flippers help it to speed through water.

The camel's broad feet, the size of dinner plates, enable it to walk on sand.

Zebras roam in open grassland where their keen sense of smell and sight warns them of predators.

Where do mammals live?

Mammals can be found almost everywhere, from the frozen wastes of the Arctic to the dry heat of a barren desert. Most live on land, but some live in water. All are well adapted for the surroundings, or habitat, in which they live.

The polar bear's thick fleecy coat protects it from the icy Arctic winds.

The jaguar's coat helps it to blend into the background in its lush rainforest home.

Amazing mammals

- Bone fossils show that mammals first lived on Earth about 200 million years ago.
- The reason mammals are *not* floppy like jelly is because they are vertebrates – animals with backbones.
- Rodents are the largest order with more than 2,200 species.

MAMMAL MEDALISTS

In a mammal olympics, the medals would go to the following: the *sloth* for being the *slowest* competitor – moving at less than 2 kph (1 mph); the *skunk* for making the *stinkiest* smell; and the *pygmy shrew* for weighing in as the *smallest* at just 2-3 g ($\frac{1}{16}$-$\frac{1}{8}$ oz).

Family life

Some mammals choose to stay in family groups, making it easier to find food, and defend themselves. Compared to other animals, mammal parents spend longer with their young.

Playtime tussles

Female lions live in permanent groups called prides and look after each other's cubs. The cubs play-fight, which is how they learn to hunt.

Meerkat watch

A gang of meerkats varies from 5-30 members. They are protective of their home, or territory, and have different roles, such as sentry duty or babysitting.

African elephants are the world's biggest land mammals.

Amazing mammals

- There are about 60,000 muscles in an elephant's trunk.
- A lion can devour 23 kg (50 lb) of meat in one meal – that's almost equal to 350 sausages.
- Ferocious fights can happen between rival meerkat gangs.

An elephant's tusks are just overgrown teeth.

Female families

Female elephants and their children stay close together in family herds. The biggest female, the matriarch, leads them wherever they go.

Different diets

What do you prefer? Vegetables, meat, fish, or a bit of everything? Mammals eat all sorts of things. They eat to produce energy, just like a car needs fuel to go.

A mixed plate
This Alaskan brown bear, like other brown bears, eats a mixed meat and plant, or omnivorous, diet. It waits to pounce on any salmon swimming upstream, but also chomps on plants, fungi, and large insects.

Keep on chewing
American bison are herbivores, which means they only eat plants. They graze on grass, and then rest. After that they chew on grass even more.

Wild mammals build their daily routine around finding enough to eat.

Make mine meat
A pack of grey wolves maul their hunting prize. As one of the world's best-known carnivores, or meat eaters, they have bodies designed for hunting other animals. They have powerful jaws and sharp teeth.

Don't stick your tongue out!

Giant anteaters wouldn't listen to this warning. They use their 60-cm (2-ft) spiky and sticky tongues to lick up termites and ants, once their clawed front feet have ripped open the nests.

The anteater pushes its long, tube-like snout into the hole.

Amazing mammals

- Wolves can eat up to 9 kg (20 lb) of meat at one meal.
- A giant anteater flicks out its tongue 150 times a minute.
- Brown bears eat a lot. The extra weight helps them survive the winter, when they sleep, or hibernate, for several months.

Moonlighters

Just as you are going to sleep, some creatures are waking up, more than ready for the night. Nocturnal mammals often have special features – such as big eyes for seeing well in the dark.

Night babies

Bush babies have large eyes for night vision and batlike ears that help them to track insect prey in the dark.

The ringtailed cat is an excellent climber and hunts in trees for small birds.

Is it really a cat?

No! The ringtailed cat is part of the raccoon family. Like the red fox, it hunts alone at night.

TALE OF ALL TAILS!

Around the world, people like telling stories about me as they can't decide if I'm cunning or clever. In Ancient Greece, Aesop wrote about me in fables. In America, I am Brer Fox who tries to outwit Brer Rabbit, and in Japan I am revered as a messenger of the Shinto rice goddess.

Night raiders

Red foxes usually hunt alone at night in woodland or open country, and increasingly in built-up areas. They will eat almost anything.

The biggest cat of all

Tigers like to hunt as the Sun sets and during moonlight hours. They are able to creep up on prey because their distinctive thick black stripes help them blend into the shadowy forests.

Cats' pupils expand to let in more light so that they can see in the dark.

On the defensive

When under attack, all mammals have ways of defending themselves. After all, none of them want to be eaten or hurt. Some will turn and run, while others use unusual methods to put off a predator.

Is it or isn't it?

Virginia opossums often play dead when under threat, hoping that the potential predator does not want to eat a dead animal! They may lie still for up to six hours.

A living ball

If threatened, the Brazilian three-banded armadillo rolls itself up into a complete ball, protecting its soft parts. Tough skin and an awkward shape prove an effective defence against most predators.

No way through

These enormous musk oxen form a defensive line or circle, if threatened by a polar bear or wolf pack. Young or weak animals are protected in the middle of the group.

Fully grown adults may leave the line to charge an attacker.

Amazing mammals

- Brazilian three-banded armadillos can curl up as soon as they are born.
- When an opossum "plays dead", its heartbeat slows down.
- If a gorilla group is threatened, they may attempt to avoid the danger by quickly heading into thick forest. This is called "silent flight".

Gentle giant?

The lowland gorilla is not an aggressive mammal, and what looks like a scary roar is actually a nervous yawn! Male gorillas will protect their social group. Defence tactics include roaring and beating their chests.

Underground, overground

Many mammals have underground homes or burrows. It's a place to have their babies or to run to from danger. Most leave their burrows to find food or water, but some, such as the mole, are true burrowers and mostly stay underground.

Underground towns
Colonies, or towns, of black-tailed prairie dogs live in tunnels under grassland that may be an incredible 5 m (16 ft) deep. The nesting chambers are lined with soft grass, and prairie dogs dig out passing places along the tunnels.

Busy burrow
Rabbits are sociable creatures and love to live in big colonies. Their burrows are large and complicated. They even have emergency exits!

Midnight feast!

These tiny wood mice are nibbling on acorns in the safety of their burrow. Their varied diet also includes berries, worms, fruits, and snails.

Tunnel vision

A pocket gopher spends most of its life underground. Its small eyes and ears, flat head, and long whiskers are all useful in a burrow.

Born to dig

You'll know there's a mole about if you see mole hills – a series of soft hills that the mole pushes up. Its huge front feet make the mole a perfect little digger.

The high life

Bats are the only flying mammals, and there are more than 1,000 species in the bat family. These furry fliers are divided into two groups: the small, mainly meat-eating microbats, and the large, mainly vegetarian megabats.

The biggest bat

The flying fox is the largest bat in the world, with a wingspan of more than 1.5 m (5 ft). It feeds on pollen and fruit.

Night fliers

A colony of Mexican fruit bats awakes at dusk, and flies off to feed on fruit and nuts. All bats are nocturnal mammals.

Roosting together

Bats often gather together in huge numbers at a single site. This may be a cave, an old building, or a hollow tree. The site must provide the bats with shelter, and protection from predators.

Finding food

Most insect-eating bats hunt using a process called echolocation. Each bat makes a series of clicks, and this sound is carried out into the air. This noise bounces off any potential prey, such as mosquitoes and moths, and sends information back to the bat. The bat can then find the prey, and enjoy its meal!

Blood sucker

This vampire bat is enjoying a tasty snack of donkey blood. Its sharp teeth easily pierce the skin, and its spit prevents the blood from clotting. Only three species of bat feed on blood.

Tent-making bats

These tiny fur balls are Honduran white bats. They only appear white under artificial light, and are well camouflaged in the murky rainforest. They create shelters from large rainforest leaves.

Primate party

Apes, monkeys, and humans are the most widely known members of a mammal group called primates. A primate party would go with a swing as most primates are playful and highly intelligent creatures.

A devoted mother

An orang-utan mother and baby stay together for about eight years. A baby clings to its mother's fur as she moves through the trees. At night, the mother makes nests from leaves for her and her baby to sleep in.

Gentle giants

Gorillas live in family groups. They are the heaviest of all primates but, despite appearances, are vegetarians. Their enthusiasm for eating forest plants and fruits can result in large potbellies.

A gripping tail

Many central or South American monkeys – such as this black howler monkey – use their grasping, or prehensile, tail as a fifth limb. With its distinctive howl, this monkey is one of the loudest of primates.

Second in class

Humans score highest for intelligence but chimpanzees are second. This chimp is using a stone as a tool for cracking open palm nuts.

If you scratch my back...

... I'll scratch yours. These baboons are checking each other's fur for ticks, or lice. It is part of a behaviour called grooming, shared by most primates. This also helps the primates to develop good friendships.

Amazing mammals

- The orang-utan's name comes from the Malay words for "man of the wood".
- Do you like pulling faces? Many primates can pull faces to show their feelings and to communicate with each other.

Amazing marsupials

Kangaroos and koalas belong to a group of mammals called marsupials. A marsupial is only partly formed when it is born and it carries on growing in a pocket on the outside of its mother's stomach, known as a pouch.

Mobile homes

A baby kangaroo, or joey, is born after just 12 days inside its mother. It crawls through its mother's fur and into a special place called a pouch. It stays in its mother's pouch, drinking her milk, for the next six months.

When the kangaroo hops, a long tail helps it to balance.

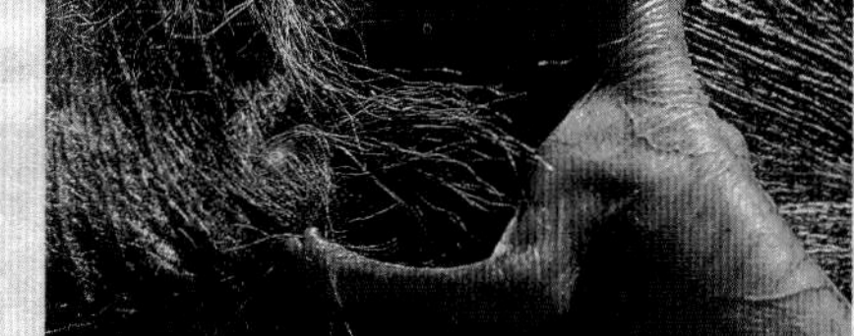

Thirsty work

A newborn kangaroo is blind, helpless, and very pink. It clings tightly to its mother's fur and will suckle continually.

Piggyback, please

A koala spends most of its life in eucalyptus trees. It sleeps for up to 18 hours a day and feeds only on eucalyptus leaves. Baby koalas live in their mother's pouch for about six months before crawling onto her back.

A fully grown kangaroo is as tall as an adult human, but at birth, it is less than 2 cm (¾ in) long.

Powerful kangaroos

The largest living marsupials, red kangaroos live in Australia. They live in groups of about 2-10 animals, with one dominant male and several females. When bounding at full speed, kangaroos can reach speeds of about 50 kph (30 mph).

WINNER TAKES THE GIRL

Male kangaroos sometimes fight over females. This fighting can take the form of "boxing". The kangaroos stand up on their hind legs and attempt to push their opponent off balance by jabbing him or locking forearms. The winner of the boxing match is the stronger male, and he gets the girl!

Deep in the jungle

Imagine a place where it rains every day but where it is uncomfortably hot and sticky. Many mammals make a good home here – among the towering trees of the rainforest.

Jungle jaguar
Hunting alone in the Amazon forests, a jaguar stalks prey on the ground, up in trees, and in water. It means its varied diet includes deer, monkeys, and turtles. Apart from man, this big cat has no rival predator.

All-day snoozer

You may not spot a sloth in the thick rainforest foliage. It moves slowly and spends most of its time sleeping. Added to that, algae grow on its coat, causing its fur to look green. This young sloth will cling to its mother for six months.

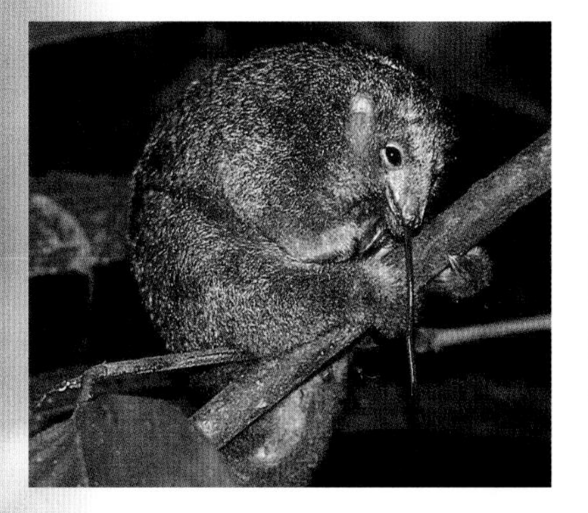

Hooked up

A silky anteater's hook-like claws are ideal for gripping a branch. It uses its red, sticky, saliva-coated tongue to scoop up ants from their nests.

Amazing mammals

- The sloth is one of the world's sleepiest mammals. It will doze in a tree for 15-18 hours a day.
- A silky anteater can eat up to 8,000 ants in one day.
- Jaguars go for a direct kill, biting through the skull of their opponent rather than seizing its neck.

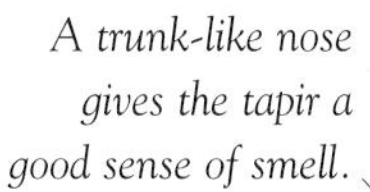

A trunk-like nose gives the tapir a good sense of smell.

In the swim

Find a rainforest river or swamp and you may spot a tapir. This timid creature likes to cool off in water, and uses it to hide from predators.

Eye eye!

Pads on the ends of a tarsier's fingers and toes help it to grip a branch while its big eyes scan the forest floor for insects to eat. Can you believe that each eye is heavier than its brain!

Forest dwellers

Unlike a rainforest, a temperate forest is ruled by the seasons – as are its mammals. In the warm spring, the young are born. In the summer, they feed and grow. Chilly autumn days see thicker coats. To survive the cold winter, some sleep, or hibernate.

What's for dinner?

One of the largest wild pigs, the wild boar spends many happy hours rooting around on the forest floor. Here it is looking for anything to eat, from roots and nuts to small animals.

Snack stop

Each autumn, red squirrels scurry about gathering nuts and pine cones. They'll store these provisions in the ground or in tree holes, raiding these "cupboards" in the winter when food is scarce.

Sleep tight

The dormouse doesn't even try to struggle through winter. It curls up, snuggles down into its leaf-and-grass nest, and sleeps, or hibernates, the winter away.

In the autumn, the antlers fall off and re-grow in the spring. They grow bigger each year.

Amazing mammals

- Dormice are named after the Latin word for sleep: *dormire*. They spend an amazing three-quarters of a year "asleep"!
- A moose has broad hooves that enable it to move through snow, muddy bogs, or lakes.
- The wild boar is the ancestor of the domestic pig.

Moosing around

The elk, or moose, is the largest member of the deer family. The male is huge. Its antlers alone can span up to 2 m (6½ ft). Forests with swampy areas provide the moose with all its food, but just imagine eating twigs or the roots of water plants!

Savanna plains

Africa's tropical grassland, or savanna, is home to spectacular groups of mammals. Life is hot and there is occasional rainfall, but grasses keep on growing for much of the year and grow back quickly after being nibbled on by the grazing herds.

Tongue stretch

Giraffes are the tallest animals in the world. Some males grow up to 5.5 m (18 ft) tall. Their long necks allow them to reach tasty leaves high up in the trees.

Predatory pride

Lions hunt many grassland mammals, even attacking young elephants and giraffes. They live in prides of 5-40 animals, made up of a few males but mainly adult females and their cubs.

All species of giraffe have different markings.

Amazing mammals

- Each zebra has a unique stripe pattern.
- Lionesses provide each other with a baby-sitting service for their cubs!
- Giraffes can gallop at speeds of 50 kph (30 mph).

Giraffe splits!

The short, wet season produces water holes that shrink as the year goes on. This drives groups, such as of giraffes, to travel great distances to find water. Their height means that they have to stretch their front legs very wide in order to drink.

Safety in numbers

Zebras often mingle with wildebeest to make a bigger herd. No-one knows why zebras have stripes, but it could be to confuse predators or for their own temperature control.

Desert homes

The Sun burns down. There is no water, and very little food. At last the Sun sets, but it is now bitterly cold. Welcome to the desert. Surprisingly, a number of mammals love it here!

Where's the water?

The spinifex hopping mouse doesn't need to drink. It gets all the moisture it needs by nibbling on plant food.

Ships of the desert

Camels are ideal desert mammals, whether they have one hump or two. They can survive for weeks and travel long distances without food or water, an ability that makes them useful for carrying things. That's why they are known as the ships of the desert.

Camels have padded feet for protection as the ground is so hot.

Cool grazers

During the heat of the day, Arabian oryx can be spotted huddled under trees. Their bright white coat reflects the light back, and therefore helps to keep them cool. There were hardly any oryx left in the 1970s, but captive breeding has meant that hundreds have been returned to the wild.

Deserts cover about 20 per cent of Earth's land.

A double row of eyelashes keeps the sand out of a camel's eyes.

THIRSTY WORK

Contrary to popular belief, a camel's hump is *not* full of water. It is actually made of fat, which the camel can live from if there is no food or water. A camel can survive for 10 months without water and then drink a *lot* very quickly – almost equal to 340 cans of drink in 10 minutes!

On the slopes

A mountain slope is a tricky place to live. The weather gets colder as you head up and the air gets thinner, with less oxygen. And there's not a lot of food! But some mammals make it their home.

A rocky home

Alpine marmots live high up in alpine meadows. If threatened, they make a loud, piercing whistle. To survive the winter, they retreat to their burrows to sleep, or hibernate, for several months.

A sky-high leap

The mountain goat is an expert rock-climber, and baby goats (kids) can walk and climb shortly after they are born. Oval-shaped hooves have a rubber-like sole that helps each goat to grip onto the slippery slopes.

Amazing mammals

- Japanese macaques are also known as "snow monkeys".
- The Alpine marmot is one of the 14 species of marmot.
- The mountain goat looks like it has a beard like other goats, but it is actually just an extension of its throat mane.

Many mountain mammals have thick fur to protect them from the icy weather.

Jump in!
These striking Japanese macaques live in the cold highlands and mountains of Japan. In winter, temperatures drop below freezing. To stay warm, the clever monkeys have learned to take a bath in the natural hot springs.

Life in the freezer

The polar regions, at the top and bottom of the world, are tough places for mammals. They need to survive the freezing temperatures, especially in winter. They also need to be cunning to find the little amount of food there is.

Boxing bunnies

Arctic hares have white winter coats to help hide them from predators. They have no trouble finding each other, though! The males box to claim a female.

Young polar bears spend up to two years with their mothers.

Polar giants

The enormous polar bear is one of the world's largest land-based carnivores. It has thick, white fur, which keeps it warm in the freezing cold, and camouflages it in the white snow. Polar bears give birth to cubs in the winter, in secure ice dens.

Made for snow

Reindeer have hollow hairs on their bodies to keep them warm, and thick furry hooves to stop them from sinking into the snow. Reindeer are migrating animals, travelling in huge herds for enormous distances in search of food.

Female polar bears nearly always give birth to twins.

Underwater world

Seals spend most of the winter in the icy water, digging air holes with their teeth in the ice above. They have thick fat (blubber) under their furry skins, which keeps them warm while they hunt fish and squid in the freezing waters.

Freshwater mammals

Fast-flowing freshwater rivers and streams, large lakes, and boggy marshes are home to all sorts of mammals. Although well adapted to water, they all have to come to the surface to breathe.

An underwater playground

Otters are well suited to water, with webbed paws, streamlined bodies, and ears and noses that close when the otter dives. These playful mammals often chase each other, and dive for rocks and shells.

A freshwater dolphin

This river dolphin lives in the Amazon river in South America. It swims through the slow-moving river channels, looking for fish and crabs. Sometimes, it swims upside-down so it can see what's happening underneath it!

Amazing mammals

- Otters and beavers belong to the Mustelidae family, which also includes mink, stoats, skunks, and badgers.
- The name hippopotamus means "river horse".
- An Amazon river dolphin has 25-30 pairs of teeth.

Lazing around...

The hippopotamus spends much of its life underwater, and can stay submerged for about 15 minutes at a time. It likes to laze in the water, with only its eyes, ears, and nose poking out.

Beavers are often thought of as pests, as they dam rivers and streams.

Beavers drag branches into place with their strong jaws.

The busy beaver

Beavers are nature's builders. They make homes called lodges out of mud and branches, complete with underwater entrances. A series of dams built around the lodge controls the flow of water, so each lodge has its own private pool.

Down by the sea

The shallow waters around the world's oceans provide a home for many mammals. Some divide their time between swimming in the sea, and breeding and caring for young on the shores. Others never leave the ocean.

Amazing mammals

- Seals, sea lions, and walruses belong to the "pinniped" family.
- The only pinnipeds able to support themselves in a semi-upright position on land are sea lions, fur seals, and walruses.
- Walruses turn pink when they leave the cold sea and their bodies warm up in the Sun.

Battle of the tusks

These large, male walruses are using their long, sharp tusks to fight over a female. Older males are often covered in scars from previous bloody battles.

The tusks of male walruses can grow to about 1 m (3¼ ft) in length.

Nearly all pinnipeds are covered in a layer of fat known as blubber.

A furry tale

Although seals appear sleek and shiny when they are under the water, they actually have two layers of soft fur. Adult fur is not as thick as baby fur, and some seal species are furrier than others. Fur keeps seals warm and is waterproof.

Dugongs were alive in the time of the dinosaurs.

A streamlined body makes all pinnipeds agile swimmers.

Underwater grazer

The dugong is also known as a "seacow", as it grazes on the seabed for sea-grass roots. This large, vegetarian mammal spends all of its life in the sea, only coming to the surface to breathe.

Swimming sea lions

Californian sea lions are fast swimmers and can move at 40 kph (25 mph) in short bursts. They can stay underwater for up to an hour, using air that is stored in their lungs.

Leap in the deep

The ocean-dwelling cetaceans are some of the most specialized mammals in the world. The cetacean family includes all whales, dolphins, and porpoises. All have streamlined bodies, can dive deeply, and hold their breath underwater for long periods of time.

Leaping high

Bottlenose dolphins are found in all of the world's oceans, except in the polar regions. Living in groups, or schools, of 4-20 animals, these playful mammals often leap above the waves.

Amazing mammals

- There are two types of whales – baleen whales, who filter food through plates in their mouths, and toothed whales, such as killer whales.
- All cetaceans breathe through nostrils on their head.
- Water supports a whale's weight; if it lived on land, it would be too big to survive.
- A humpback whale calf can grow until it is approximately 16 m (50 ft) long.

Breaching giants

This humpback whale is leaping high out of the water. This leaping is known as breaching. All whales breach, and we don't really know why they do this. It may be to warn off other whales, communicate with their group, or just for fun. A whale this enormous will make a huge splash when it hits the water.

Water baby

Humpback whales tend to have their calves in the spring, in warm, tropical waters. The calf is born tail first, and its mother helps it to the surface so it can breathe. The calf will stay with its mother for about a year.

Friend or foe?

Like any big family, sometimes everyone gets along, sometimes they don't! Humans can have close or useful relationships with other mammals. There are also times when mammals can cause problems.

A powerful snow sledge

These dogs are Alaskan malamutes. Their strength, long legs, and thick coats make them ideal for pulling sledges over ice. They are devoted to their owners.

Pesty nibblers

Some people keep mice as pets, but others squeal and are scared at the sight of them. House mice can cause havoc as they spoil food if they get a chance to nibble, and they can eat through books and wires and spread disease.

Amazing mammals

- In New Zealand, sheep outnumber people by 20 to 1.
- To survive, mice breed quickly. A house mouse can have up to 70 baby mice in a year.
- Pet kittens can look after themselves at about eight weeks, whereas cubs from big cats rely on mum for much longer!

Popular pets

Years ago, wild cats killed mice and rats that ate people's grain. Humans began to care for the friendly cats and soon they became pets.

A sheep haircut

Sheep were first used, or domesticated, by people many years ago for their wool, meat, and milk. Sheep are shorn once a year when their coat is at its thickest.

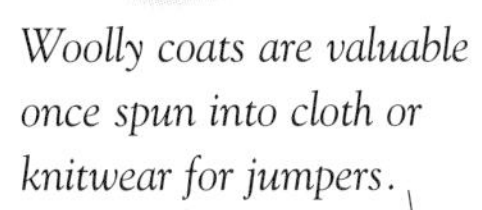

Woolly coats are valuable once spun into cloth or knitwear for jumpers.

Under threat

Can you imagine having your home taken away and being hunted? Many mammals live with these threats all the time.

A watchful eye

Just as famous people have bodyguards, rhinoceroses have guards to protect them from poachers.

This black rhinoceros's horn can grow up to 1.4 m (4½ ft).

Shocking news

Rhinos are under threat because people kill them for their horns. The horns are used for making traditional Asian medicines and dagger handles.

A park for everyone

Large herds of bison wander and graze in safety in America's Yellowstone National Park because nobody is allowed to shoot them. They share the park with many other mammals, including grizzly bears.

Amazing mammals

- One hundred years ago, there were a million black rhinos. Now, less than 5,000 are left. Numbers are increasing, though.
- The Chimfunshi orphanage in Zambia looks after 100 chimpanzees.

Pressures for pandas

A giant panda needs to munch on a variety of bamboo stems every day. However, habitat changes mean there aren't enough types of bamboo plants left. Reserves in China are trying to help.

Orphaned chimps quickly learn to feed from a bottle.

Feeding time

Many people want to help sick, hurt, or unwanted mammals. These chimpanzees are cared for in an orphanage in Zambia.

Polar journey

A reindeer herd is migrating in search of food. Help the herd reach its destination by answering these questions correctly.

START

In the winter, polar bears give birth to cubs in...
See page 34

the open

water

secure ice dens

Young polar bears stay with their mothers for up to...
See page 34

four years

two years

one year

Alaskan malamutes have long legs and...
See page 42

rough coats

thick coats

thin coats

find food

Reindeer have hollow hairs for...
See page 35
camouflage
sucking up water
warmth
Seals spend most of the winter...
See page 35
in the icy water
hibernating
migrating
White winter coats help Arctic hares to...
See page 34
box each other
hide from predators
FINISH

True or false?

Find out if you know these mammals well. Spot if these statements are true or false.

All species of **bat** feed on blood.
See page 19

Giant pandas feed on tree leaves.
See page 45

When an **opossum** "plays dead", its heartbeat slows down.
See page 15

Japanese macaques live in cold highlands and mountains.
See page 33

A black **rhinoceros's horn** can grow up to 1 m (3 ¼ ft).
See page 44

The **tusks** of a male walrus can grow to about 2 m (6½ ft).
See page 38
A dormouse **hibernates** during the winter.
See page 27
An **orang-utan** mother and baby stay together for about a year.
See page 20
Camels have either one or two **humps**.
See page 30
Huge front feet make a mole a perfect digger.
See page 17
American bison are herbivores.
See page 10

Jungle safari

Have you ever been to a jungle? Play this game with your friends to experience what the jungle has to offer.

How to play
This game is for up to four players.
Move down
Move up
You will need
A dice
Counters – one for each player
Trace over the jeep outlines, or cut and colour your own from card. Each player takes turns to throw the dice, and begins from the START box. Follow the squares with each roll of the dice. If you land on an instruction, make sure you do as it says. Good luck!
Frightened by a gang of meerkats. Move forward 3
Stopped to admire zebras. Miss a go
Wildebeest running towards you. Miss a go
Chased by a colony of bats. Move forward 5
Rest a while. Miss a go
Run along with a wild boar. Move forward 3
START

Who am I?

Take a look at these close-ups of mammals in the book, and see if you can identify them. The clues should help you!

- I am born after just 12 days inside my mother.
- After that, I stay in my mother's pouch for six months.

See page 22

- Using my sticky, 60-cm- (2-ft-) long tongue, I lick up termites and ants.
- I have clawed front feet.

See page 11

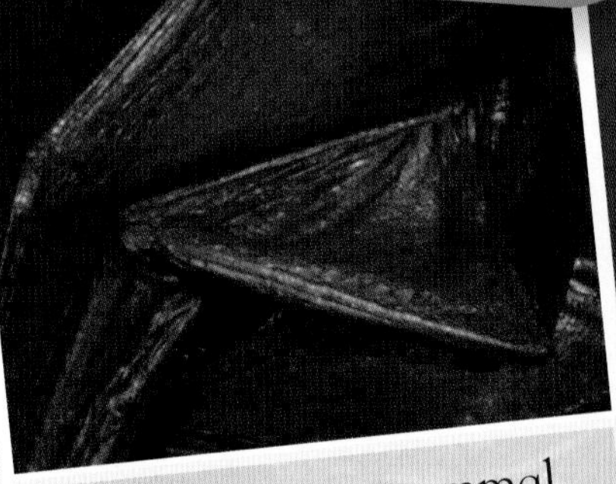

- I am a flying mammal, with a wing span of more than 1.5 m (5 ft).
- I feed on pollen and fruit.

See page 18

- I am an excellent climber.
- I hunt alone in trees for small birds at night.

See page 12

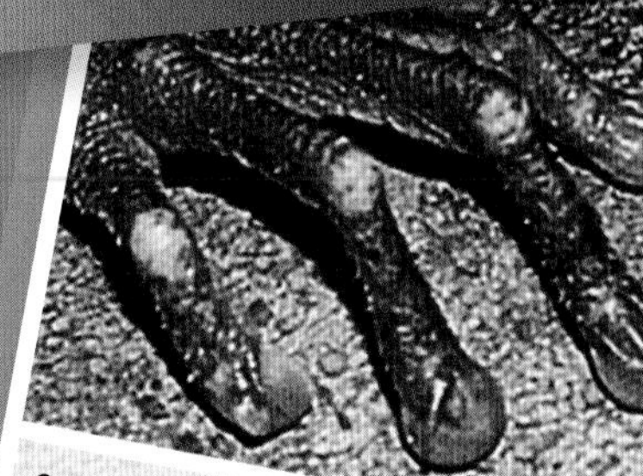

- Pads on the ends of my fingers and toes help me to grip branches.
- I have big eyes, each heavier than my brain.

See page 25

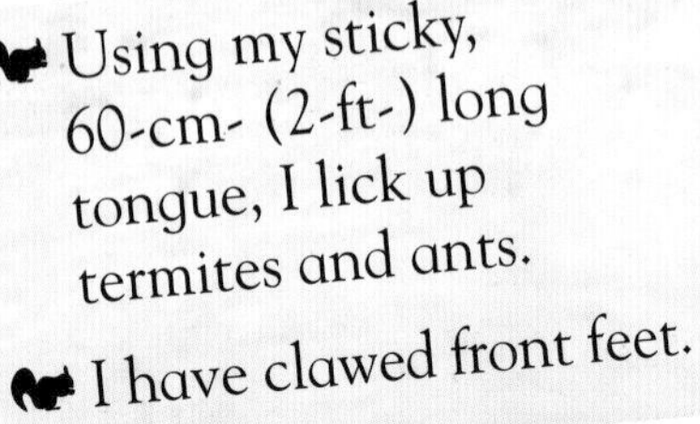

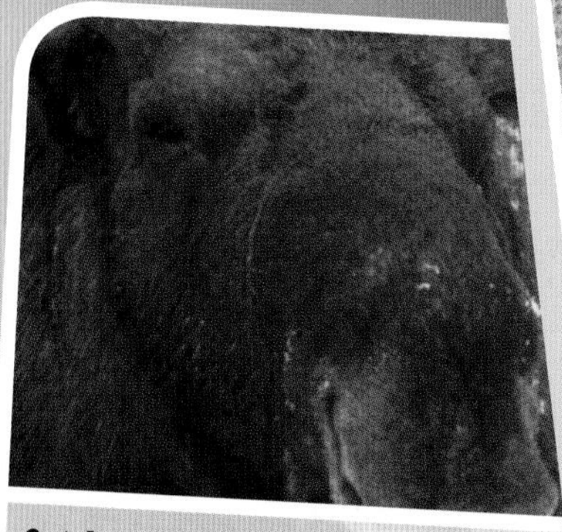

My antlers can span up to 2 m (6½ ft).

In autumn, the antlers fall off. They re-grow in the spring.

See page 27

Most of my life is spent in eucalyptus trees.

I sleep for up to 18 hours a day.

See page 23

If threatened, I roll myself up into a complete ball.

My tough skin protects me from predators.

See page 14

I am not an aggressive mammal.

My defence tactics include roaring and beating my chest.

See page 15

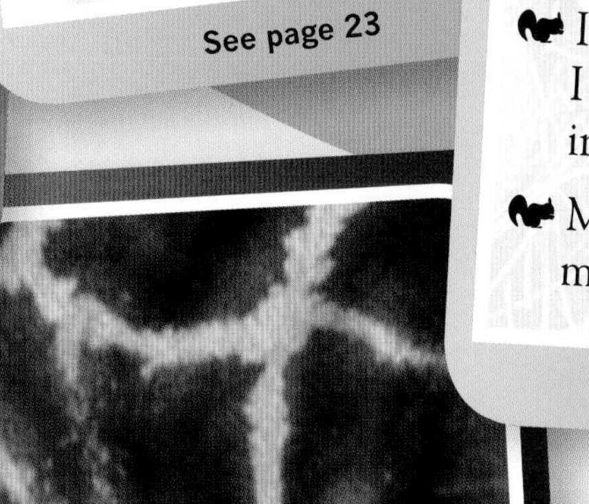

We are the world's tallest animals.

Our long necks allow us to reach leaves high up in the trees.

See page 28

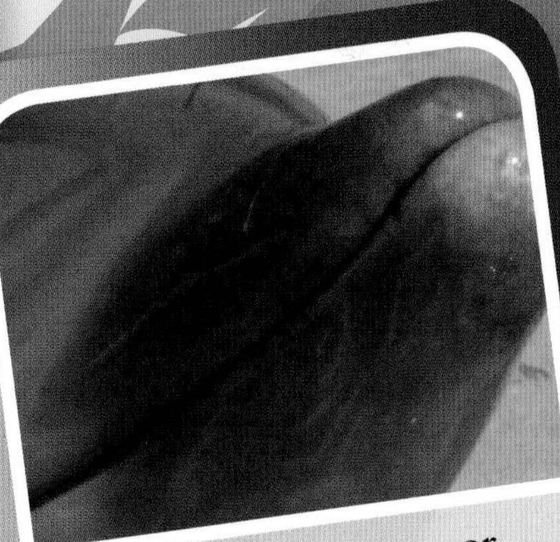

We live in groups, or schools, of 4-20 animals.

We are playful mammals, often leaping above the waves.

See page 40

Glossary

Here are the meanings of some words that are useful to know when learning about mammals.

Camouflage a colour or pattern that matches an animal's surroundings and helps disguise it.

Carnivore a meat eater.

Colony a group of animals that live together.

Echolocation a means of using echoes to steer an animal towards food and to build a picture of what is around them.

Habitat the place where a creature or plant naturally lives or grows.

Herbivore a plant eater.

Hibernate describes the period some animals spend asleep in the winter.

Marsupial a mammal whose young is born undeveloped. It continues its development in a pouch.

Migration a long journey which some animals undergo each year to find better living conditions.

Nocturnal active at night.

Omnivore a plant and meat eater.

Predator an animal that hunts other animals for food.

Prey an animal hunted for food.

Prehensile tail a tail that can grasp (like a hand).

Stalk to approach prey quietly, so that it does not notice.

Suckle the means by which a baby mammal feeds from its mother.

Warm-blooded able to maintain a constant, warm body temperature, despite the surrounding temperature.

Animal alphabet

Each mammal featured in this book is listed here, along with its page number and which area it comes from.

Index

Acknowledgements

Dorling Kindersley would like to thank:
Beehive Illustrations (Andy Cooke) for original illustrations; Rose Horridge and Charlotte Oster for picture library service.

Picture credits:

The publisher would like to thank the following for their kind permission to reproduce their photographs:
a=above; c=centre; b=below; l=left; r=right; t=top

Alamy.com: Steve Bloom Images 8-9; 21tl. **Ardea London Ltd:** Ian Beames 27tr; Liz Bomford 12cr; Thomas Dressler 28tl; Jean-Paul Ferrero 20; Ferrero - Labat 3; Chris Knights 43br; Charles McDougal 13; Stefan Meyers 43tl; S. Roberts 19t; Adrian Warren 19cr. **Bruce Coleman Ltd:** Bruce Coleman Inc 17tr, 24; Peter A. Hinchliffe 17tl; Werner Layer 27tl; Orion Press 6-7; Hans Reinhard 26; Pacific Stock 40; Staffan Widstrand 25tr; Gunter Ziesler 25cl. **Corbis:** Tom Brakefield 11; Bryn Cotton / Assignments Photographers 36bl; Michael and Patricia Fogden 25cb; Gallo Images 1, 4cr, 9tr, 28bl; Layne Kennedy 10cr; W. Wayne Lockwood, M. D. 32-33; Steve Kaufman 33br; W. Perry Conway 37br; Paul A. Souders 10tl; Kennan Ward 27br; Tom Brakefield 52cl; image100 49br; Kennan Ward 53tl. **Dorling Kindersley:** University Museum of Zoology, Cambridge 49tr, 51cla, 59bl. **Dreamstime.com:** Antonella865 50cr; Rinus Baak 48bl; Nilanjan Bhattacharya 59cl; Driescronje 51c; Iakov Filimonov 46clb; Gatito33 58tl; Jon Glover 50bl; Isselee 46cb; Jeanninebryan 59cra; Martha Marks 59cla; Mgkuijpers 49tl; Gentoo Multimedia 47c; Roman Murushkin 59tl; Pascalou95 58clb; Scattoselvaggio 47bl; Sdbower 59tr; Nico Smit 59br; Toneimage 58cl; Vasiliy Vishnevskiy 58br; Suzanne Weegar 48crb; Robin Winkelman 53bl; Simone Winkler 47tr. **Philip Dowell:** 6cb, 7cb. **The Image Bank / Getty Images:** Joseph Van Os 10bl; Art Wolfe 29. **ImageState:** 18l, 41b; Natural Selection Inc 44. **FLPA - Images of Nature:** Foto Natura Stock 39tl; David Hosking 45tl; Gerard Lacz 22-23, 42; Minden Pictures 22bc, 37tl, 39, 45tr; Mark Newman 37c, 38cl; W. Wisniewski 54. **FLPA:** Mitsuaki Iwago / Minden Pictures 52tr. **Fotolia:** Nadezhda Bolotina 46tr; Ewan Chesser 58cra; frank11 51tc; Eric Isselee 53tc, 58tc, 58crb; Valeriy Kalyuzhnyy / StarJumper 50c, 50cb, 51tl, 51tr, 51cl; wojciech nowak 51bl. **Nature Picture Library Ltd:** Mark Payne-Gill 14cl, 31tr; Mark Payne-Gill 52bl, 53cra. **Natural Visions:** Andrew Henley 16tc. **N.H.P.A.:** Ant Photo Library 30tr, 38b; Pete Atkinson 6cra; Mark Bowler 21tr; Martin Harvey 7clb, 22bl, 44tr, 45bl; Daniel Heuclin 25br; Rich Kirchner 7cla; Alberto Nardi 6crb; Rod Planck 56l, r; Jonathan and Angela Scott 8tl; Karl Switak 16bl; David Woodfall 36bkgd. **Oxford Scientific Films:** Clive Bromhall 21cl; Mike & Elvan Habicht 14tr; Mike Hill 4tl; Peter Lillie 21br; Partridge Films Ltd 12tl. **The Photographers' Library:** L & D Jacobs 4-5b. **Photoshot:** NHPA 52br; **Powerstock Photolibrary:** 32tr; Brandon Cole 41tr. **Science Photo Library:** Gregory Dimijian 19bl; Phil Dotson 12cl; Adam Jones 5tr; Stephen J. Krasemann 18bkgd; Renee Lynn 28c; Phil Dotson 52cb, 53tr. **Still Pictures:** Klein/Hubert 23tl. **Stone / Getty Images:** Daniel J. Cox 34-35, 35br; Paul Harris 35tl; David Myers 14b; Art Wolfe 34tr. **Telegraph Colour Library / Getty Images:** John Downer 30-31; Tony Evans - Timelapse Library Ltd 16-17; Gary Randall 15. **Jerry Young:** 7ca.

All other images: © Dorling Kindersley.
For further information see: www.dkimages.com